"AUTOPSY"

AN AUTOPSY KNOWN AS

THE POST-

MORTEM EXAMINATION

NECROPSY,

THE AUTOPSIA

CADAVERUM OR THE

OBDUCTION IS THE

SPECIALIZED

SURGICAL PROCEDURE DONE

THROUGH THE

ESAMINATION OF A

CORPSE TO DETERMINE

THE CAUSE

AND MANNER OF DEATH

AND ALSO

TO EVALUATE ANY

DISEASE OR INJURY THAT

MAY BE PRESENT.

THE PATHOLOGIST
PERFORMS ALL
AUTOPSY PROCEDURES.
AUTOPSIES
ARE DONE FOR EITHER
LEGAL OR MEDICAL
PURPOSES. A FORENSIC
AUTOPSY BRINGS THE
CAUSE OF
DEATH TO BE A CRIMINAL
MATTER WITH THE
CLINICAL OR ACADEMIC

AUTOPSY PERFORMED TO GET OR FIND THE CAUSE OF DEATH. IF IT IS THERE BROUGHT FOR UNKNOWN OR UNCERTAIN DEATH OR SUCH ETC,

RESEARCH PURPOSES AN AUTOPSY CAN BE FURTHER CLASSIFIED INTO THE EXTERNAL EXAMINATION SUFFICES AND DISSECTED INTERNAL

EXAMINATION AS CONDUCTED.

THE NEXT OF KIN MAY BE REQUIRED

FOR INTERNAL AUTOPSY IN THE

SAME CASES. ONCE THE AUTOPSY IS

DONE THE BODY IS RECONSITIUTED

BY SEWING IT BACK TOGETHER.

THE MAIN GOAL AND THE MAIN

PRINCIPLE IS TO

DETERMINE THE CAUSE OF

DEATH.

THE AUTOPSY STATES

WETHER ANY

MEDICAL DIAGNOSIS AND

TREATMENT BEFORE

DEATH WAS

APPROPIATE. AN AUTOPSY

ALSO

CAN BE CARRIED OUT FOR THE

PURPOSE OF MEDICAL RESEARCH.

AUTOPSIES ARE USED IN CLINICAL

MEDICINE FOR THE IDENTIFICATION

OF MEDICAL ERROR. A SYSTEMATIC

REVIEW OF STUDIES OF THE

AUTOPSY CALCULATED THAT IN

ABOUT 25% OF AUTOPSY A MAJOR

DIAGNOSTIC ERROR WILL BE FOUND

AND REVEALED. A PARTICULAR

STUDY FOUND THAT OUT OF 694

DIAGNOSES "AUTOPSIES REVEALED

171 MISSED DIAGNOSES, INCLUDING

21 CACERS, 21 STROKES, 11

MYOCARDIAL INFRACTIONS, 10

PLUMANARY EMBLOI AND 9

ENDOCAROLITIS, AMONG

OTHERS"

THERE ARE FOUR MAIN
TYPES OF
AUTOPSIES; MEDICO-
LEGAL OR
FORENSIC OR CORONER'S
AUTOPSIES TO FIND THE
CAUSE OF
DEATH AND IDENTIFY THE
DECENDENT. SECOND ONE
IS
CLINICAL OR
PATHOLOGICAL

AUTOPSIES DONE TO

DIAGNOSE A

PARTICULAR DISEASE

AND

RESEARCH PURPOSES.

THE THIRD

ONE IS ANETOMICAL OR

ACADEMIC

AUTOPSIES PERFORMED

BY STUDIES

OF ANATOMY STUDENTS

OF

ANATOMY FOR STUDY

PURPOSES

ONLY. AND FOURTH IS

VIRTUAL AND

MEDICAL IMAGING

AUTOPSIES

PERFORMED BY

UTLILIZING IMAGING

TECHNOLOGY ONLY.

THE FORENSIC AUTOPSY

CONSISTS

OF FOUR MAJOR PARTS;

NATURAL,

ACCIDENT, SUICIDE

DEATHS OR

UNDETERMINED.

THE UNDETERMINED

DEATHS

INCLUDE DEATHS SUCH

AS THE

DEATHS IN ABSENTIA

MEANING

DEATHS AT SEA OR
MISSING
PERSONS. CLINICAL
AUTOPSY HAS
TWO MAJOR PARTS
PURPOSES
ONE IS TO GAIN MORE
INSIGHT INTO
PATHOLOGICAL AND
SECOND TO
DETERMINE THE FACTORS
OF DEATH.

THE EXTERNAL
EXAMINATION
FIRST AFTER THE BODY IS
RECEIVED
FOR AUTOPSY IT IS
PHOTOGRAPHED
OF HAIR AND NAILS AND
LIKE ARE
TAKEN, THEN ALSOTHE
BODY MAY
BE RADIOGRAPHICALLY
IMAGED.

THEN, ONCE THE EXTENAL EVIDENCE

IS COLLECTED THE BODY IS

REMOVED FROM THE BAG AND ANY

WOUNDS FOUND ARE EXAMINED,

THE BODY IS CLEANED, WEIGHTED,

AND MEASURED IN PREPARATION

FOR INTERNAL
EXAMINATION.

THE GENERAL
DESCRIPTION OF THE
ETHNICITY, SEX, HAIR
COLOR, EYE
COLOR, DISTINGUISHING
FEATURES,
BIRTHMARKS, SCARS,
AND MOLES,

ETC, IS MADE. A VOICE RECORDER IS

NORMALLY USED TO RECORD THIS

INFORMATION. THE INTERNAL

EXAMINATION, THE BODY IS PLACED

UNDER THE BODY CHEST UPWARD.

THE INTERNAL EXAINATION

CONSISTS OF INSPECTING THE

INTERNAL ORGANS OF THE BODY

TO EVIDENCE OF TRAUMA OR ANY

OTHER INDICATIONS WITH THE

CAUSE OF DEATH. THE RECONSTITUTION OF THE BODY, THE

AUTOPSY OF THE BODY

SUCH AS

THE FOLLOWING

PROCEDURE

EXAMINATION. AFTER

FULL

EXAMINATION THE BODY

HAS AN

EMPTY CHEST CAVITY, IT

FLAPS

OPEN, ON BOTH SIDES,

AND THE

SKULL FLAPS ARE
PULLED OVER
FACE AND NECK. IT IS
VERY USUAL
TO OPERATE THE FACE,
ARMS,
HANDS, LEGS
INTERNALLY.
THE TERM AUTOPSY
DERIVED FROM
ANCIENT GREEK
"AUTOPSIA"

MEANING "TO SEE FOR
ONESELF"

THIS AUTOPSY HAS BEEN
USED

EVERSINCE THE 17TH
CENTURY

REFERING TO THE

EXAMINATION

OF THE INSIDE OF THE

HUMAN BODY

TO DISCOVER ANY

DISEASE AND

CAUSE OF DEATH.

AROUND 3000 BC

THE ANCIENT EGYPTIANS

WERE THE

FIRST CIVILIZATION TO

PRACTICE

THE REMOVAL AND

EXAMINATION

OF INTERNAL ORGANS OF

HUMANS.

IT WS THE RELIGIOUS

PRACTICE

KNOWN AS THE

MUMMIFICATION.

AUTOPSIES WHICH

OPENED THE

BODY TO DETERMINE THE

CAUSE

OF DEATH WERE ATLEAST

IN THE

EARLY THIRD MILLENIUM.

ALTHOUGH WERE

OPPOSED MANY

ANCIENT SOCIETIES
WHERE IT WAS
BELIEVED THAT AN
OUTWARD
DISFIGURMENT OF THE
DEAD
PRESENTED FROM THE
ENTERING
AFTERLIFE. ALSO THE
POST-
MORTEM EXAMINATION IS
FOR

MORE COMMON IN
VETERINARY
MEDICINE THAN IN HUMAN
MEDICINE.

FOR MANY SPECIES THAT
EXHIBIT

FEW EXTERNAL

SYMPTOMS OR

THAT SUITED TO CLINICAL

EXAMINATION WHICH IS A

COMMON

METHOD USED B

VETERINARIANS TO

COME TO A DIAGNOSIS.

A PHYSICIAN CANNOT

ORDER AN

AUTOPSY WITHOUT THE

CONSENT

OF THE NEXT OF KIN. A

MEDICAL

EXAMINER CAN ORDER AN

AUTOPSY

WITHOUT THE CONSENT

OF THE

NEXT OF KIN. THE DEATHS

THAT ARE

INVESTIGATED BY THE

MEDICAL

EXAMINER OR CORONER

INCLUDE

ALL SUSPICIOUS DEATHS

AND THEN

DEPENDING UNPON

JURISTICTIONS,

IT MAY INCLUDE DEATHS OF

PERSONS NOT BEING TREATED BY A PHYSICIAN FOR A KNOWN MEDICAL CARE FORLESS THAN24 HOURS OR DEATHS THAT OCCOURED DURING OPERATIONS OR OTHER

MEDICAL PROCEDURES. IN ALL

OTHER SUCH CASES A CONSENT

MUST BE OBTAINED FROM THE NEXT

OF KIN BEFORE THE AUTOPSY IS

PERFORMED, EVEN AT ACADEMIC

INSTITUTIONS OR HOSPITALS, THE

NEXT OF KIN HAS ALL THE

RIGHTS

TO CONTROL THE SCOPE

OF THE

AUTOPSY FOR EXAMPLE;

EXCLUDING

THE BRAIN FROM

EVALUATION OR

LIMITING THE PROCEDURE

OF THE

ABDOMEN. MEDICINE

FUNDING

REINBURSMENT FOR
AUTOPSY

IS THEOROGICALLY
INCLUDED IN THE
FIXED PAMENTS FOR
HOSPITALS.
THESE FUNDS NOT
TYPICALLY
ACCEPTED FOR
AUTOPSIES, THE

PRACTICE PAY AUTOPSY ARE

DISTORT. AUTOPSIES ON ANIMALS

WERE PERFORMED NOT FOR THE

STUDY FOR DISEASE BUT RATHER

FOR THE PRACTICE OF PREDICTING

THE FUTURE AND COMMUNICATING

WITH DIVINE FORCES.
FAMILY
MEMBERS MAY CONSIDER
AN
AUTOPSY WHEN A
MEDICAL
CONDITION HAS NOT BEEN
PREVIOUSLY DIAGNOSED
OR IF
THERE ARE QUESTIONS
ABOUT AN

UNEXPECTED DEATH THAT

APPEARS

DUE TO NATURAL CAUSES

OR IF

THERE AR GENETIC

DISEASES OR

CONDIDTION THAT ALSO

MAY BE THE

RISK FOR DEVELOPING

AND WHEN

THE DEATH OCCOURS

UNEXPECTED

DURING MEDICAL OR
DENTAL

PROCEDURES OR WHEN
THE CAUSE

OF DEATH COULD AFFECT
THE LEGAL

MATTERS AND WHEN THE
DEATH

OCCOURS DURING THE

EXPERIMENTAL

TREATMENT.

AN AUTOPSY MAY ALSO BE

REQUIRED IN DEATHS THAT HAVE

MEDICAL AND LEGAL ISSUES AND

THAT MUST BE INVESTIGATED BY

THE MEDICAL EXAMINER OR

CORONER'S OFFICE, THE

GOVERNMENTAL OFFICE ATLEAST IS

RESPOSIBLE FOR INVESTIGATING

DEATHS THAT ARE IMPORTANT TO

THE PUBLICS HEALTH AND WELFARE. DEATHS MUST BE

REPORTED TO AND INVESTIGATED BY

THE MEDICAL EXAMINER'S OR

CORONER'S OFFICE AND CAN VARY

BY STATE AND MAY INCLUDE THOSE

THAT HAVE OCCOURED; SUDDENLY

OR UNEXPECTEDLY, AS A RESULT

OR ANY TYPE OF INJURY, UNDER A

SUSPICIOUS CIRCUMSTANCES, UNDER OTHER CIRCUMSTANCES, DEFINED BY LAW. IN SOM EOF THESE DEATHS AN AUTOPSY MAY BE REQUIRED. THE MEDICAL EXAMINER HAS LEGAL AUTHORITY

TO ORDER AN AUTOPSY WITHOUT
THE CONSENT OF THE DECEASED
PERSON'S FAMIL NEXT OF KIN.

IF AUTOPSY IS NOT REQUIRED BY
LAW IT CANNOT BE PERFORMED
UNLESS THE DECEASED PERSON'S

FAMILY GIVES
PERMISSION. AN

AUTOPSY IS ESPECIALLY

IMPORTANT DONE TO;
DETERMINE

AS POSSIBLE WHAT
CAUSED THE

DEATH AND CONFIRM OR
EXCLUDE

DISEASE DIAGNOSIS MADE
BEFORE

DEATH AND DOCUMENT THE

PRESENCE OF A DISEASE THAT WAS

UNDIAGNOSED BEFORE DEATH THEN

ALSO COLLECT SAMPLES OF BODY

FLUIDS OR TISSUES FOR POSSIBLE

GENETIC STUDY AND THEN COLLECT

EVIDENCE AND
INFORMATION IN
CRIMINAL CASES. ALSO
TO HELP
HEALTH DEPARTMENTS
OR
GOVERNMENTAL
AGIENCIES
IDENTIFY AND TRACK A
DISEASE OR
POTENTIAL PUBLIC
HEALTH

HAZARD.

THE END